PHYSICS AND THE ALCHEMY OF CONSCIOUSNESS

Also by Pir Vilayat Inayat Khan
SAMADHI WITH OPEN EYES

Physics and the Alchemy of Consciousness

Edited from a talk given by Pir Vilayat Inayat Khan
to a holistic health conference in San Diego, California,
on September 4, 1977

SUFI ORDER PUBLICATIONS
P.O. Box 396
New Lebanon, New York 12125

Printed in the United States of America
ISBN 0-930872-05-3

Physics and the Alchemy of Consciousness

Pir Vilayat Inayat Khan

Sufi Order Publications
NEW LEBANON, NEW YORK
1978

Physics and the Alchemy of Consciousness

Specialists are said to know more and more about less and less, while the jack-of-all-trades knows less and less about more and more. The hope is that the current trend towards the holistic view will make those influencing human health bridge the gap between the two. Those involved in that trend are responding to an uncanny hunch that holistic health is the next step towards the unknown azimuth of human progress. We may be unwittingly or discerningly abreast with current scientific perspectives, such as the hadron shoe strap theory, which stress the interaction, interdependence and interpenetration of all phenomena in the universe.

> The parts of the whole have to find their place with the other parts of the whole organism which they compose fully.
>
> —Fritjof Capra

Following the lead set by Einstein in bridging the electromagnetic and gravitational laws, it is not surprising that in this decade we are researching the confluence of physis and psyche, or better, the congruence. We use the word congruence because an infallible sense of purposefulness seems to be compulsively spurring human realization onwards to an ever more encompassing grasp of the implicit programming underlying all phenomena.

In the field traditionally known as 'esoteric' our objective is working with the psyches of people first, and then on further levels. We begin by liberating the minds of men from the preconceived ideas (mind-binds) that throttle our impelling need for meaningfulness. Obviously, everyone sees things from his vantage point; consequently, an opinion can never be absolute. Next we try to help people over those hurdles which might be called heart-binds. Have you ever realized how lucky it is that what you wished for ten years ago never materialized; or, perhaps, have you been sorry that it did?

Then we aim at liberating people from their rather random self-image by encouraging them to grasp the bounty of inherited qualities. Some of these qualities are recessive because they have been displaced by the dominant qualities fostered by the contingencies of upbringing or the prevailing circumstances and environment. After that, we tend to strengthen the person's self-confidence, unearth unknown talents and idiosyncrasies, and foster growth. We are working with the emotions of people, inculcating in them that psychic energy that triggers off transformation or transmutation by means of that spring behind all birth and death, which is ecstasy. One cannot live on bread alone; one needs wine. (I am, of course, misquoting.) This higher dimension of emotion is associated with a sense of glorification which arises out of one's awareness of communicating with the vastness or splendor or meaningfulness beyond the narrow range of ordinary perception and understanding. Here are the emotions that wipe out hatred, violence and crime.

Basically, we are working with the parameters of consciousness, modulating its focus. From being entrapped in the environment, consciousness spreads into cosmic dimensions, ultimately grasping the archetype beyond the exemplar: roseness beyond the rose. Consciousness is ever outreaching its polarization in our sense of I-ness by intuiting an overall consciousness beyond what we think of as our consciousness.

The methods used to present these objectives involve, of necessity, physiological and consequently physical factors. Up to the present, *gurus* have simply applied time-honored pragmatic techniques which work. Most fanciful innovations or adaptations eventually reveal an unforeseen flaw. For example, *kundalini*, which is part of a way of life intended for *sannyasins* living in the wilderness, if practiced in an apartment in New York by a well-intended but self-styled *guru*, may end up in the psychiatric ward. Most traditional *gurus* do not like people to mix their practices with those of other *gurus*, just as doctors do not like patients to follow the prescriptions of two doctors simultaneously. It takes a lot of knowing to probe the wisdom of the methods. But the validity of the methods is evidenced by their extraordinary results. For example, some *yogis* gain control of previously autonomic functions of the body, including brain waves. The focus of consciousness can then offset the balance between the anabolic and catabolic rate, and so on.

We are living at a time when we are able to ascertain and even measure so-called subjective experiences of meditators that had not heretofore been corroborated by others. For example, the Kirlian photographs show emanations from the fingers and the entire aura. Bentov confirms that by the sheer fact of thinking of light, people's auras burn brighter. This is reassuring for *gurus* under fire from laymen who think they are promoting hallucinations in their pupils. Generally, here is a pointer to the use of creative imagination in spiritual techniques.

Such corroborations from scientists embolden *gurus* to pursue the cultivation of the audition of the so-called abstract sound (*saut-i-sarmad* of the Sufis, *shabd* of the Hindus, or the sound current, promoted by the late Sant Kirpal Singh). This is the same as the Pythagorean "symphony of the spheres." Claiming such paraphysical auditions still exposes people to the threat of being committed to a mental institution. Admittedly, one must be able to discriminate between the so-called subjective (or rather, reflexive) messages which project unconscious urges and the genuine enhanced auditory acuity of trained esotericists who have extended their sense of perception beyond the normal auditory frequency range. This is rather similar to the distinction between intuition and fantasy.

We are working with sound in the *mantra* (the Sufis call it *wazifa*). The plexuses of the autonomic nervous system (for example, the solar plexus or the cardiac plexus) respond to being bombarded with sound, like a bell struck, and will produce specific sound frequencies regardless of the nature of the object that sets them into resonance. In fact, all objects in the universe are in a state of resonance, and this resonance can be enhanced to the point of making it audible to human ears. What is more, it creates a sympathetic consonance in other objects in the universe. By placing vowels and consonants in the appropriate plexus, while bringing out the overtones, one triggers off a transformation in character. But what is more interesting still is that ideas or archetypal principles have their own resonance, which is a key to primeval languages (as opposed to conventional languages).

The meditator is taught to work with light as a musician works with sound. As mentioned, one can heighten one's auric radiation by an activated consciousness. There are techniques of raising one's temperature by burning intensely, and then transmuting the energy into auric light. One can work on one's glance by setting one's sight at infinity with closed eyes and then maintaining this focus when the eyes are opened. This technique can be used to see auras. Further, there is a technique of revitalizing one's aura by looking into the sun. Beginners are taught to shield their retina against direct exposure, and gradually to increase the exposure. Assuredly, some of these techniques could be used in a kind of light therapy. These are the practices that are traditionally used to attain illumination.

One of the direst problems encountered by people involved in action and emotional stress is when exhaustion strikes the magnetic field. In some cases the body is fairly unscathed. Here, medical intervention proves ineffective, unless the doctor can convince the patient to rest, giving him a chance to escape the environment creating the stress. However, oftener than not, the patient carries his stress with him on holiday. We deal with this problem on several levels. We teach meditators to become aware of their magnetic fields (or life fields, researched by Harold Burr with a vacuum tube voltmeter). It is our experience that gaining consciousness of the life field enhances its intensity. We also experience an extension of the field as we exhale physically, and a contraction, together with an increase of pressure, as we inhale. The focus of consciousness equally influences the life field. Expanding the span of consciousness, for example, by thinking of the vastness of the starry sky and consciously drawing energy from sidereal space, will intensify the field. We can become aware of charging our field from the magnetic field of the earth, and of ejecting polluted energy back into the earth's field. We can even transmute physical energy into magnetic or astral, auric or akashic or even beyond, as in *kundalini* techniques. We welcome a confirmation on these subjective impressions by scientists in their laboratories, hoping it will help us better to understand what we are promoting in others, and thus to improve our techniques. This energy, so enhanced, may be applied to healing by laying on of hands, through the breath, through the glance, or even by thought at a distance. It is like a blood transfusion, for life kindles life, passing from a more highly charged battery to a less charged one.

Many people in our time feel a need to call a halt to the pressure of everyday obligations and to attempt to gain a bird's eye view of their problems and idiosyncrasies. To answer this need, we have come forward with a retreat program. Normally, one fails to see the woods for the trees. The retreat gives people an opportunity to experience those states that one sometimes intuits but never has time to clinch. This can trigger off a much longed-for transformation in one's being. We offer people involved in everyday living the (albeit short-lived) opportunity to experience something of the inner condition and attunement of the recluse. This means utter seclusion for a period of time. However, so much know-how goes into conducting an individual retreat that we put candidates through a six-day preparation course followed by six-day conducted group retreats. Under these circumstances, the meditators can be expected to complete an optimal and safe retreat. The stress upon people unused to being thrown upon their own devices is so great that we feel compelled to take every precaution to insure that they are strengthened by the experience rather than maladjusted or mentally impaired, as could well happen unless the retreat is carefully monitored. A guide is assigned to each meditator who checks up on him regularly.

The overall schedule is based upon the stages in the process of transformation outlined in the *ars regia*, the royal art of the alchemists. This method was used by the Sufis, who were among the first alchemists, outlined by Valentinus, and has been referred to in our time by Carl Jung. If "as above, so below," human personalities follow the same basic changes as minerals in the womb of the earth or elements in the Mendeleev periodic table. Unless our personalities undergo a process of disintegration, we remain unchanged. Only glibly self-opinionated beings would accept such a state. It would cut out that perspective of hope which is the spring of all life. However, unless the process of disintegration of the psyche is completed by a rebuilding process, the human being runs the risk of schizophrenia, paranoia, and other personality and neurological disturbances. On the other hand, it is a great way to be free from the tyranny of the self.

Since the personality is an expression of the notion of the self, the personality undergoes change. It must necessarily do so, just as the body does. One has to learn to transfer one's notion of oneself from the fixed image one has acquired to the notion of being continuity in change, a current rather than an entity. This is again corroborated by the new vistas in science: there are no divisible entities in the universe.

> The particle can no longer be seen as a static object, but as dynamic patterns.
>
> —Paul Dirac

As in the *Avatamsaka-sutra*, "the solid outlines of individuality melt away."

People panic when they try to hold on their erstwhile personality as it fails them, like the ground pulverizing under their feet. We have to learn, as St. Francis said, to pass through a stage of being nothing. Somehow, growing up means unmasking a score of illusions and giving up one's crutches. Even facing pain is a prerequisite for growing, but the Sufis affirm that our illusions are a protection that must be removed gradually. If one were to realize how things really are, one would be shattered irretrievably. One would realize that one's mind is unable to cope with one's realization, as St. John of the Cross points out in the 'dark night of understanding.' The process of parturition from the matrix of the ego may need a midwife or a doctor. Practices are given which involve recollecting retrospectively the events of one's life, as in a film. Simultaneously, one sees the transformation of one's personality which results from the events, and sometimes influences them as well. Occasionally, causal sequences emerge clearly in one's awareness.

The second stage consists in disidentifying oneself from everything that is subject to becoming and decay. By identifying with the eternal archetype of one's being, one salvages that which survives death and disintegration. Did not Christ beckon us to cleave to the eternal life? This is the transcendental dimension; one might call this process discovering a vertical continuity. An example is astral travel: one thought one was a body, and now one realizes that one is the same person, yet so different in appearance and feeling. Better still, one can identify with the archetype of one's personality. One discovers oneself as always having existed as a particular blend of qualities, and thus one is able to observe the personality one assumes in one's present existence as an offshoot of one root, which has sprouted every now and again in different incarnations in the course of history.

The technique used is *kundalini* (or ascending-descending breath). As one breathes in, one transfers one's attention from one *chakra* to the next, along the spine. (For those unfamiliar with this Eastern term, *chakra,* one might say that most *chakras* correspond to the plexuses of the autonomic nervous system, although they are traditionally considered to be centers belonging to the higher bodies.) Here is a case of the application of consciousness to enhance a neurological process: namely, the flow of energy across the synapses. Conversely, there is the effect of the physical forces, including attendant secretions, upon the setting or focus of consciousness. As each *chakra* corresponds to a sphere or body, one identifies progressively with: the magnetic field, the astral body, the body of *prana* (the Hindu term for life force, at a level I have not yet ascertained with certainty), the light aura, the akashic body (or one's vibrational network), the third eye (which is like a built-in radar set), and finally transcendental consciousness or intelligence. According to esotericists, not only is the latent energy in each center released to awaken the next one in a chain reaction, but there is a transmutation of energy from matter to spirit in slow stages, which is what is meant by resurrection.

The third stage, which marks the apex of the *solvae* phase, constitutes the final leap in the vertical continuity from the personal to the impersonal. "More and more all-inclusive and impersonal": this phrase of Sri Aurobindo corresponds to the present view of science that particles are processes rather than objects. According to current scientific thinking, the universe is looked upon as indivisible, so that the word individual has lost its meaning (unless one has grasped that each one contains all the others; at least this is true virtually, in the DNA structure of life cells).

First one experiences oneself as pure consciousness: not *a* consciousness, but consciousness, and beyond consciousness, pure intelligence, since "intelligence becomes consciousness when faced with an object" (Hazrat Inayat Khan). Then one experiences oneself as life force—not *a* life force—and beyond life force, pure spirit: the life of life, as the Sufis say, or we could say the catalyst of incipient life. In addition, spirit is envisioned as the thrust which breaks the sclerosis which sets in when life forces crisscross, stalemate, and become cyclic. Spirit primes a greater leap, as, for example, from the water state to the vapor state, or the shift of an electron from one orbit within the atom to another. More generally, one could cite all the observations in application of the quantum theory. Mystics experience sudden changes authenticated by what is called initiation; Zen teachers call these changes *satori*.

The more we try to ascertain what we mean by energy, the more elusive it becomes. Psi has been admitted into the arcana of plausible forms of energy, and astral projectors are moving feathers in hermetically closed spaces with their astral bodies' kinetic force. We may finally give access to the Holy Spirit as a valid force experienced by mystics in meditation and used in healing more effectively than the life field. St. John of the Cross and the Sufi martyr al Hallaj experienced a divine action upon them, which is precisely the meaning of the *significato passiva* of the early Church Fathers.

Then one experiences oneself beyond one's existence in time and space. One may have noticed all along that there was a transformation in the sense of time. One's notion of time is a measure of one's inability to embrace reality in one take. It is like the cinecameraman who is obliged to pan, whereas a camera with a panoramic lens may shoot less and less of more and more, and, at close range, more and more of less and less. For the worm, the blade of grass he crossed a minute ago belongs to the past; for the eagle, the hill he flew over perpendicularly twenty minutes ago is still there. The eagle's notion of the present is wider. From the top of the hill, the interconnection between events is easier to grasp; therefore, meditators seek to attune their consciousness to a transcendental dimension where one can grasp the cause behind the cause behind the cause, rather than be drawn into a plethora of accumulated, apparently unrelated facts. When all things have fallen out of focus and only the transcendental cause stands out, that is called *samadhi*.

This is best illustrated by the technique of *nidra yoga*, which consists in maintaining the continuity of consciousness during sleep. Most people have the impression that they awake into day consciousness, but the meditator practicing *nidra yoga* observes that the dreamer has the impression of having awakened out of day consciousness and experiences himself as a much more universal being than he seems in his day consciousness. One can monitor dreams by presuggestion: as a therapy, this has been used successfully by Dr. Ross of Syracuse in the treatment of epilepsy. There is a whole field of research in harnessing the autocreative faculties. There is creative imagination in dreams, which sets in at a level above the regurgitation of the day's impressions or the projection in symbols of the inhibited or frustrated motivations of the unconscious. Moreover, the *nidra yogi* gains the master key to the mystery of man when he penetrates the normally hermetically sealed state of deep sleep. This state is invaluable in recuperating from the expenditure of energy during the day's activity. Dreaming performs a function in integrating mental and psychic conflicts in the disturbed, but is unnecessarily tiring for the clear personality. Therefore, bypassing the dream period and becoming aware of the advantage of prolonging deep sleep is the clue to sleeping 'faster' as well as reaching transcendental dimensions of experience.

The *Mandukya Upanishad*, the origin of Vedanta, draws a parallel between the stages experienced in sleep and those of meditation. It is remarkable to note that the highest state aimed at in *yoga* meditation presents a convincing number of factors in common with deep sleep: there are no images (no rapid eye movements), the heartbeat is slowed down and so the pulse, while brainwaves recorded on the EEG produce theta and delta frequencies. The advantage of *nidra yoga* is that the meditator can carry back his impressions back into physical life, and thus permeate it with celestial influence. Swami Rama of the Himalayan Institute and I were able to produce theta and delta waves while maintaining a remote awareness of the environment. One remembers having had a body that is located somewhere on the planet, and one used to be able to use one's mind to channel thought, but this seems now like a hoax. Logic seems to be a framework for consciousness trapped in its focus in the notion of the self, limiting understanding, whereas one intuits transcendental logic such as that referred to by Ouspensky in the *New Model of the Universe.* A meditation theme devised to promote *samadhi* describes an immaculate state at the source of life, illustrated by a landscape of the soul of snow and ice, where a clear stream runs in a bed of crystal, scintillating in diaphanous light. This scene is often associated with the Virgin Mary.

At this point ends the first phase of the alchemical process, including the first three stages, the lunar phase, including the minor mysteries.

In the fourth phase, the person on retreat proceeds to rebuild his personality intentionally and consciously, participating in all the phases of his own recreation; it is like experiencing one's birth all over again. This is called the alchemical marriage. No doubt the miracle of life results from the fusion of earth and heaven, the convergence, or rather congruence in a point of space and time of the whole universe, as Teilhard de Chardin says.

The subject is made to earmark the qualities he inherits from his father and mother, resolving conflicts that may have existed between them. Then he includes ancestors, wherever known, envisioned as numberless strands that have become intertwined, nay interfused, into his being. This includes inheritance from the animal, vegetable and mineral realsm, and the inheritance from the djinn and angelic spheres and from all planes, and finally the divine inheritance. These are the attributes we previously acquired as eternal archetypes before we co-opted the substance of the different planes in our descent. Christ beckoned us to be aware of our inheritance: "Be perfect, as your Father in heaven is perfect." That is, identify with the divine perfection, rather than with the exemplification of it in the personality. Should one be aware of one's divine inheritance, one could rightfully claim to be the son.

In the fifth stage the candidate has to learn to identify himself with his new image and resist the influence of friends who tend to force him back into the image that they have become used to. A practice borrowed from the East consists in representing oneself as sitting in front of an ideal person—the *guru* image of one's most utopian daydreaming—and gradually identifying with him until one realizes that one could not have imagined him if one did not have his idiosyncrasies embedded in one's own psyche. It is easier to see oneself in another self who can manifest what one is better than one can oneself. In addition, one uses the techniques of the *mantram* whereby one visualizes a quality or attribute in its perfection (for example, divine power or joy or peace), letting the impression sink deeply into one's being. This illustrates Poincare's assertion that man is endowed with a sense of infinity and eternity: he can always imagine a larger number, he can conceive a further star than any star earmarked, or a time earlier or later than any conceivable moment in time. The meditator is thus taught to observe his qualities as the exemplification of the attributes, and thus to channel them into reality.

Returning to life in the sixth stage, the meditator is now equipped with insight gained at high altitudes, and is expected to resist the tendency to slip back into his personal vantage point. He is therefore supposed to maintain the supreme perspective while involved in life. Shankaracharya tells of a man who in the dark thought he saw a snake. In the daylight, he saw it was a rope. Thus he knew, the next night, that what looked like a snake was really a rope, but it really did look like a snake. This is called by the Hindus *samadhi* with open eyes, or *sahaj samadhi*. It is like a double focus, or, better still, stereoscopic vision. One realizes that people see things from the vantage point of the planet earth. According to this vantage point, the sun turns around the earth. But when one awakens beyond this vantage point, as Copernicus did, one visualizes the earth revolving around the sun. This is the key to relativity.

It is a matter of being continually aware with a high degree of acuity, so that one never loses sight of the fact that one has landed on earth at some point in time and fashioned the fabric of the planet, or rather, focused and organized the life forces of the universe, into that marvelous instrument, the body. One performs this creative act in order to be able to experience the conditions of the planet, but also in order to fertilize the planet with a new dispensation of intelligence that one has channeled down from the heavens. Alchemists called it the spiritualization of matter and the materialization of spirit. One would like to think that this impression of meditators confirms the discoveries of the conversion of mass into energy and energy into mass.

This way of looking at things is further developed in the *sattipatana* or *vipassna* techniques of meditation in Buddhism. One observes one's body so objectively that one wonders at the miracle of being able to make this package of the fabric of the planet walk. One realizes that what one assumed were one's thoughts were just the way the mind works. One watches oneself react to situations and realizes that one has let oneself believe that one was being involved, whereas all that is actually there is a highly conditioned and automated function of the conjunction of tendencies and attitudes reflecting the nature of the universe. But the greatest breakthrough is reached when the meditator watches his consciousness experiencing objects or situations, as from a second level of consciousness. Actually, the sense of 'I' has systematically retracted from all that one had originally identified with. One has been gradually eliminating the notion of the personal subject which, as in the principle of Heisenberg, masks the actual phenomena. The drawback of this method is that many, failing to identify with the impersonal consciousness, simply indulge in introspection, which becomes tedious and contradicts the very objective aimed at.

The fallacy in *samadhi* with open eyes is that, basically, one is looking at things from a vantage point which grasps the cause behind the cause, whereas really to account for the way things happen, one has to consider the object—teleology, rather than ontology—evidencing free will beyond law. Hazrat Inayat Khan says laws of nature are habits that have sclerosed the divine action into the process of becoming. If we exercise our real free will beyond the conditioning of body and mind, we monitor the divine will which is continually overcoming stale laws by spontaneous creativity. The Sufis call this the divine intention, which is continually new. Things happen not only because of what we did in the past (*karma*), but also so that we might become what we would be if we could be what we should be.

Our main thrust hits deeply into the whole quest for the unfoldment of the personality. This is a feature of the metaphysics of the Sufis, who value what is achieved in existence. The objective of the contemplatives in earlier times was liberation from existence and the self. This new orientation of the Sufis is emphasized in Hazrat Inayat Khan's teaching on the art of personality and God's fulfillment in existence in and through man. Suppose you consider the personalities of men as plants, and their eternal archetypes as the seeds. If there seems to be anything gained by existence, by the cross-pollination of the personalities of creatures, it is that these personalities acquire qualities which, though latent in their seeds, are nevertheless fulfilled by being thus enriched. If the process of resurrection is like salvaging the perfume of the flower from the death of the petals, then the sphere of archetypes will be enriched by the interchange of archetypes taking place on earth between their embodiments in creatures.

Thus the Sufis foster a form of consciousness which I have called omega consciousness, which embodies Teilhard de Chardin's concept of a convergence or interpenetration of the consciousness of all beings as its final objective. Certainly this is the next step in evolution. It is Martin Buber's I-Thou relationship, instead of the I-it relationship of ego-centered society. Omega consciousness starts by experiencing what it is like to be another person, getting into the consciousness and feelings of another, and finally reaches the divine consciousness, not in the night of time before creation, but in its immanence within creation. This is planetary consciousness, as Hazrat Inayat Khan called it. The leaf becomes increasingly aware of the consciousness of the tree in joy and pain. Nicholaev was channeling this collective level of consciousness when picking up Kamensky's telepathic messages. We are now becoming aware of a group alarm signal system linking all beings in their involvement in the whole. Individual identity stands in the way of operating at this level. The reason why dervishes and rishis and anchorites generally are endowed with such uncanny intuition is that they have lost their individual identity and are functioning at this level.

The day we are faced with the problem of communicating with beings on other planets, we shall be forced into thinking in terms of our identity as planetarians, beyond our individual identity. Then we will realize that our language, the tool of our individual thinking, stands in the way of our intuition. The only language that will enable us to communicate will be intuition. Similarly, when communicating with beings on planes other than the physical, we shall have to overcome our consciousness of ourselves as created beings in order to communicate at the only level that links all beings: the divine consciousness.

The purpose of man is like the horizon: the farther he advances, the farther it recedes. There comes a time when one discovers a cosmic purpose beyond one's personal purpose. This realization was outlined by Hazrat Inayat Khan as the motto of the New Age: the awakening of humanity to the divinity of man. God bless you.